The Executive's Cybersecurity Advisor
Gain Critical Business Insight in Minutes

Mike Gable

120

Se

Methods Press

2021

Founded in 2021, Se Methods Press makes complicated technical subjects easy to understand. These subjects include tech sales, sales engineering, IT professional services, and Cybersecurity. The views expressed in its publications are entirely those of the author and do not necessarily reflect the views of the staff, or owners of Se Methods Press.

http://www.semethods.com

The Executive's Cybersecurity Advisor

Se Methods Press, An Imprint of Journeyman Publishing LLC
12524 Wheatgrass Ct, Parrish, Florida 34219 USA

Errata: Errors@journeymanpublishing.com

First Edition, 2021. Cataloging-in-Publication Data is available from the Library of Congress.

ISBN: 978-0-9885402-3-1 (printed)
ISBN: 978-0-9885402-4-8 (e-book)
ISBN: 978-0-9885402-5-5 (hardback)

Journeyman Publishing is a speciality publisher for Cybersecurity, IT Professional Services, Software Performance Engineering, Technical Selling, & Software Quality Assurance books. If you have an interest in becoming a published writer, please contact us at:

111 Foster Mill Circle, Pauline South Carolina 29374 USA
newauthor@journeymanpublishing.com

For My Sharon

About the Author

Mike Gable was born in Barranquilla, Colombia and spent his formative years in Latin America. While at Wake Forest University, he declined an invitation to act with a traveling theater company. After college, Mike worked in software testing and wrote patches. A friend (now his wife) suggested he interview for a Sales Engineer position because sales presentations combine his love of tech with his passion for acting. That SE job started a career where Mike has led every customer-facing team, from sales to technical support.

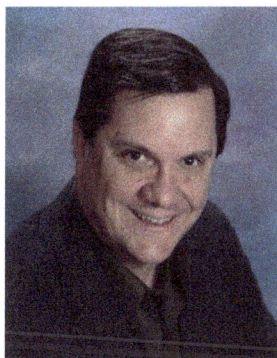

Mike is a Cybersecurity expert and award-winning presenter. He is best known for his storytelling, humor, and for his ability to explain the most complex subjects in simple terms.

Contents

Executive Summary

If you're concerned about spending a fortune on Information Security (Infosec), without any idea what your organization is paying for, this book is for you. Cybersecurity is easy to understand once you know the basics, starting with the big picture:

- As you read this, cybercriminals are attacking every organization, including yours.

- No solution or combination of solutions will stop every attack.

- Noticing successful attacks and stopping them as quickly as possible reduces their costs[1] – possibly all the way to zero.

- Cybercriminals have two goals. disruption and theft. They are trying to shut down your business, steal your data, or both.

---

[1]Hard costs include fines, ransom, and legal claims. Soft costs include loss of brand equity, loss of customers, and opportunity costs.

Effective Cybersecurity defense is about applying three disciplines to your Information Technology assets:

1. **Prevention** – Stop as many attacks as possible

2. **Detection** – Detect the attacks Prevention missed

3. **Response** – Undo the effects of successful attacks

Today, Prevention and Detection are not 100% effective. Some attacks are not stopped by Prevention measures and go undetected. These attacks succeed; they steal data and disrupt organizations. The fact that attacks succeed makes Response the most business-critical discipline. Prevention and Detection also play crucial roles; you want the best measures you can afford. But you need to operate with the assumption that Prevention and Detection will eventually fail. And when they fail, your organization needs a Response plan.

A CEO told me he was concerned about his organization falling victim to Ransomware attacks, since a number of companies have paid millions in Cyber-ransom. He asked his Cybersecurity leader, "What are we doing about Ransomware?" The answer was basically, "Don't worry. We have great prevention measures in place." That answer is likely accurate, but it was the wrong question. The question, "What will we do if a Ransomware attack shuts down our organization?" gets to the heart of the issue. Even the best Prevention measures fail, and some attacks move so fast there is little time to detect them and limit their damage.

There are hundreds of Cybersecurity vendors with thousands of competing solutions. This book is not about selecting any particular solution. Instead, it explains the three Core Cybersecurity Disciplines (Prevention, Detection, and Response) and shows how they apply to each information asset type – all without incomprehensible jargon and unexplained acronyms. It then provides questions and talking points for conversations with Cybersecurity leaders that will provide clarity into how they direct your organization's Cybersecurity budget.

Learn everything you need in the following pages or skip to Chapter 6 and dive into a conversation with your Infosec leader. When you understand the basics, Cybersecurity is manageable.

2

Introduction & Primer

This book provides non-technical business leaders with the information needed to understand, review, and approve Cybersecurity investments. To the uninitiated, Cyber is an incomprehensible labyrinth of arcane and expensive solutions, all somehow working to prevent costly breaches.

Introduction

Every month or so, talk of a major breach hits every news outlet. It seems that even organizations with the biggest budgets and the best talent are powerless to stop attackers. In reality, Infosec teams stop attacks every day, but attacks keep coming as attackers get more motivated and sophisticated. Some say Cybersecurlty is simply about keeping all the windows and doors to our information systems locked. The statement is true, but it ignores complexity and scale. There are millions of windows and doors in the typical organization, some of them move while others appear and disappear. Cybersecurity practitioners work to find and lock every door and window – a huge job. Attackers only need to find one unlocked door, or one lock they can break before getting caught.

I've met with CEOs, CFOs, and CAEs who wished their Infosec leaders could speak in business terms. As one CFO put it, "They always want more money and can't justify it except to say we're at risk without the added spend. When will Cyber be cost-effective?" Some say we need a new generation of Infosec leaders with expertise in finance and risk management. These hybrid *Business plus Cyber* leaders may be the future, but any business leader can communicate with today's Infosec leaders on their terms, provided they understand the fundamentals explained in this book.

The methodology described herein is my creation. It relies on familiar concepts, but it will be new to many Cybersecurity practitioners. Cyber evolved without an accepted overall strategy. It is a collection of disparate technical fields born from the need to protect IT assets. To someone outside Cyber, Network Security and Workstation Security appear similar. To practitioners, those two disciplines are as different as Neurosurgery and Italian cooking. It is up to security leaders to meld their disparate groups into one team focused on breach prevention. The simplified methodology used in this book will not only help business leaders communicate with their Infosec counterparts, it may also help organizations improve their overall security postures.

I use these terms in the following chapters:

- **Attackers** – Individuals or groups who perpetrate cyberattacks

- **Cloud** – Infrastructure and servers owned by someone else that you can pay to use over the Internet

- **Cyber** – Short for 'Cybersecurity'

- **Infosec** – Information Security – The teams who defend against Cyber attacks

- **Malware** – Attacker software, including viruses, Ransomware, and spyware

- **Vulnerability** – A flaw that can be exploited by attackers

The word *breach* seems to be developing two different definitions. In the press, a *breach* is a costly and often embarrassing event where business is disrupted, or data is stolen (or both). To Infosec practitioners, a *breach* is a common occurrence where an attacker defeats an Infosec defense. I recently spoke with a CISO (Chief Information Security Officer) who complained that his Board of Directors wanted a solution that would, "stop all breaches." While it is not clear which definition the Board used, only the press definition is a conceivable, albeit ambitious, goal today.

Cybersecurity Primer

There are millions of cybercriminals, ranging from casual individuals to well-funded syndicates and nation-states. Thanks to the Internet, they can all access your assets without visiting your location. It is a statistical certainty that at any given time attackers are testing your defenses, whether you are an individual on a laptop computer or the world's largest bank.

Data Theft

The purpose of Cybersecurity is to prevent data theft and business disruption. Data is stolen for resale, industrial espionage, or publication. Some cybercriminals steal any information with the hope they can then sell it in criminal marketplaces. Any information stored on a computer could be sold to your competition, used by criminals, or made public.

Business Disruption

Business disruption takes several forms. Years ago, Denial of Service (or Distributed Denial of Service, DDOS) attacks were common. In these attacks, cybercriminals simulate large groups of users to consume all website or application capacity, creating a service outage. There are many effective defenses against these types of attacks, but experts believe their popularity waned because they were not profitable for attackers.

Ransomware

Ransomware, first used in 1989, is now the most popular type of business disruption. In a Ransomware attack, malicious programs encrypt[2] your files, rendering them useless, crippling or shutting down operations. The criminals then offer to sell you a password to restore your files. To avoid paying the attackers, you could restore your files from a backup (assuming the backup is unaffected), or you may be able to get the password from a Cybersecurity vendor (some publish passwords for widespread attacks). As a final option, you could pay the attackers. There is, of course, no guarantee the attackers will provide the password – or that they even have it. Some attackers have no intention of providing the passwords and therefore have no reason to keep them. US law enforcement originally recommended against paying ransomware attackers, but they have changed their recommendation a few times in the last few years.

Unlike data theft, where the attacker's profit depends on the value of the stolen data to the criminal marketplace, Ransomware provides attackers with several illicit revenue opportunities. The most obvious is an extortion payment for restoring your operation; you determine the value instead of the criminal marketplace. In some cases, Ransomware attackers also steal your information and extort additional funds not to publicize or sell it. These revenue opportunities, coupled with the near-total anonymity of cybercurrencies, make Ransomware a very enticing criminal business model.

While most Ransomware attacks are extortion attempts, the goal of some attacks is disruption. If this is the case, paying the ransom accomplishes nothing – other than enriching the criminals. Some experts believe that Ransomware for disruption is part of 21st Century warfare and that some of what we're seeing today are test attacks by mercenary syndicates and nation-states.

---

[2]Unencrypted files can be read by applications written to handle the file format. For example, unencrypted Portable Document Format (PDF) files can be read by dozens of applications. Encryption uses mathematical formulas to scramble the document making it illegible. Most applications recognize that a file is encrypted and ask for the password to decrypt it, making it legible again. Most encryptions can be decoded without the password but doing so requires specialized applications can take weeks or months.

Attacker Motivation

Cybercriminals are most often motivated by financial gain, but this is not always the case. There are unconfirmed reports that developing nations have cybercriminals steal Intellectual Property from industry-leading organizations to give their nascent industries competitive advantages. Criminal syndicates have claimed that for a fee, they will steal any information or disrupt a competitor or adversary. More troubling are reports that nation-states are looking to disrupt key supply chains or vital infrastructure components. Keep these possibilities in mind if you find yourself negotiating a ransom payment.

Testing Infosec

Infosec defenses are rarely tested by anyone except criminal attackers. An army without drills would likely lose a lot of battles, just as a football team without scrimmages would likely lose a lot of games. It is remarkable that Infosec is successful when many teams are never tested by a friendly adversary. A few Infosec teams have in-house adversaries that test Cyber defenses, and some hire *White Hat Hackers*[3] to do the same. Most organizations rely only on real attackers to test their defenses. Compare this approach with finance, where public companies have internal and external auditors test every vital accounting process. Remember, Cyber is maturing after only 30 years while modern finance has been around since the Renaissance.

---

[3]White Hat Hackers, also called Ethical Hackers, are experts in Penetration (often abbreviated Pen) Testing. Organizations hire them to test Cyber defenses, just as some organizations hire Private Investigators to test physical security.

3

Core Cyber Disciplines

The world of Fire Safety, like other pursuits focused on reducing negative outcomes, condenses into three disciplines:

- Prevention

- Detection

- Response

On the **Prevention** front, Fire Safety has led to changes in building materials, fabrics, and home electrical installations. These preventive measures have saved countless lives, but they are not 100% effective; buildings and homes still burn. When **Prevention** fails, we count on early **Detection**: smoke detectors, heat sensors, and fire alarms. Early **Detection** is of no use without a **Response**. To that end, we have automated sprinkler systems and Fire Departments.

Cybersecurity relies on the same Core Disciplines. Infosec organizations invest in **Prevention**, **Detection**, and **Response**. **Prevention** measures stop most attacks. **Detection** measures notify administrators about attacks that Prevention missed. These notifications, called *alerts*, are handled by a **Response** team or individual. They verify the alerts are not false alarms and return the affected systems to their pre-attack states.

Core Cybersecurity Disciplines

- **Prevention** – Stop as many attacks as possible

- **Detection** – Detect the attacks Prevention missed

- **Response** – Undo the effects of successful attacks

Cyber Disciplines Overview

All three Core Cyber Disciplines are required for effective Cybersecurity today. In the early years of Infosec, little attention was paid to Detection and Response. Organizations believed Prevention solutions were so effective that damaging breaches were almost impossible. I worked for a leading vendor at that time, and I remember speaking with an Infosec leader who told me, "We don't plan for [breaches]. We pay you to stop them."

Today, we know that current Prevention measures do not stop every attack; virtually every Infosec organization invests in Detection and Response as well as Prevention. **Response is the most business-critical of the three Cyber Disciplines** because, unlike Infosec leaders in 2004, we understand that the issue isn't *if* your organization will be the victim of an attack, but *when?*

Ransomware attacks (explained in Chapter 2) illustrate the importance of Response. These attacks spread very quickly and may be noticed by users well before any Detection measures. If a Ransomware attack gets through an organization's Prevention measures, it could render hundreds or thousands of workstations and servers useless in minutes. A Response plan that restores those systems in a few hours is nothing short of heroic.

Prevention measures are required because without them the number of successful attacks could make Response impossible. If at some point in the future, a Prevention measure reaches 100% effectiveness, the assets it protects will not need Detection or

Response (at least in theory). If someone built apartments using methods they claimed were *100% Fireproof*, would you live there without smoke detectors or sprinkler systems? I wouldn't; I'm afraid arsonists would see *100% Fireproof* as a challenge.

Cybercriminals have two goals: data theft and disruption. Detection is vital in data theft because most attempts are stealthy, and instances go undiscovered for months or years. A large-scale theft – a few million patient records, for example – could take days or weeks to complete. Cybercriminals copy the information out slowly to reduce their chances of being detected. Response also relies on Detection. You can't respond to an attack you don't know about, just as the Fire Department won't come to your house if they don't know it is burning.

Response

Response is the most business-critical discipline because today's Prevention measures cannot stop every attack. Even if you purchased every prevention solution (cost-prohibitive for most) and deployed all of them (hurting performance and reliability) and disconnected from the Internet, you could still fall victim to an attack. The chance of a successful attack is predicated upon two things: How many attackers your organization attracts and the quality of your Prevention measures. Every organization attracts some attackers. Large organizations have data that criminals can sell for significant sums of money, but smaller organizations are also attractive targets. Ransomware costs attackers very little, and it allows them to target several organizations at the same time. A cybercriminal could disrupt ten small organizations and earn more in ransom than a single large organization might pay.

Whether there is a dedicated Incident Response Team (IR or IRT), or if Response is handled by an Infosec generalist, the job is the same: confirm the Detection (alert) was not a false alarm – and that the asset is indeed compromised – and return it to its previous, uncorrupted state.

Response is ideally completed quickly enough to minimize business disruption and to prevent attacks from spreading. In theory, the Response Teams provide feedback to Prevention to stop future attack variants.

The most time-consuming part of Response is remediation, returning systems to their pre-breach, uncorrupted states. Malware removal tools for workloads and workstations work quickly, but they are typically only used for emergencies or low-priority systems. These tools can miss malware components, making reinfection likely. Because of this uncertainty, most Infosec teams rebuild infected systems, or restore them from known-good backups. Manual rebuilds and restores can take a lot of time, so Infosec practitioners often rely on desktop management solutions to speed things up. These solutions have become integral to Response in many organizations, but there can be political conflicts over their use as they are usually owned by Information Technology (IT), not by Infosec.

Breaches today are inevitable; having Response plans to undo their damage quickly can be the difference between a tough day at the office and an extinction-level event. Ideally, these plans are tested, because as Moltke (the Elder) wrote, "No plan survives contact with a hostile force." Let's walk through a hypothetical scenario.

1. You are CEO of a company with 300 employees, 10 in-house Linux servers, and 300 employee workstations, all running Windows.

2. One Monday morning your desktop computer has a surprise for you. You try and open your nearly completed annual report and find that Word won't open it.

3. You then get a pop-up on your screen. It says your files have been encrypted – and asks for \$1M in Bitcoin in exchange for the passwords needed to decrypt the files.

4. You soon discover every employee has the same problem.

5. It's official: you are a Ransomware victim.

6. The files on all 300 workstations are encrypted. The Linux servers are unaffected.

7. Your Infosec leader arrives, and she says, "We just tested our recovery plan. We'll restore all the systems and we'll be back in business in no time."

8. It isn't quite that simple, of course. Many Ransomware attacks infect target systems days or weeks before encrypting them. A recent backup could be infected. Older backups, say from 90 days ago, are likely clean – but many users could lose months of work.

9. Your Infosec leader was prepared for this eventuality. She has a test-bed system (called a *sandbox*) where she can safely observe the malware that encrypted everyone's files.

10. She configures the file restore process to avoid restoring the malware files and everyone is back up by lunch.

11. She has a bit more work to do, but you are back in business.

This Response example has a happy ending, but it could have been a more prolonged outage. The outage was short-lived because the Infosec leader had a plan in place and was prepared for a Ransomware attack.

Some attackers also encrypt system backups. In those cases, restoring the files may not be an option. A Ransomware Response plan should take that possibility into account.

Detection

Detection is vital because it triggers Response. Attacks are often stealthy and go undetected for months or years. A large bank – that at the time was spending over $400M on Infosec – fell victim to an attacker that stole sensitive information. The bank's world class Infosec team did not detect the attack. A good Samaritan found the information in a public forum on the Internet and called the bank's customer service number. I can only imagine the reaction of the executives who spent tens of millions of dollars on Detection solutions only to have this damaging breach detected by an unpaid volunteer.

Even if the Detection source was an unpaid volunteer, a Response plan should be put into action as soon as the attack is detected.

Anecdotes like the bank's failure to detect their data loss are fueling innovation and growth in Cyber Detection. At its core, Detection is about two conflicting ideas:

- Detect every Cyber-related anomaly

- Prioritize the most critical items

Today, while most organizations still do not detect every anomaly, many are generating thousands of alerts each day – far more than their Response Teams can handle. Many of the alerts are false alarms, and a single attack can generate thousands of alerts. Dealing with very large alert volumes can lead to *Alert Fatigue*[4]. Detection is clearly not where we need it to be today, but it can be valuable even in its current state. Some specific alerts are well-known early attack indicators. Getting those alerts to Response Teams quickly has turned some potentially massive breaches into non-events.

Think of Detection solutions as Cyber smoke alarms. Like early smoke alarms, Detection solutions generate a lot of false positives. They are annoying, but they can be effective. Unlike a home with 4 or 5 smoke alarms, large organizations have hundreds or thousands of Cyber smoke alarms generating hundreds of thousands (possibly millions) of alerts each day. Determining which of these alerts warrants immediate attention is presently a top issue in Cyber.

Most organizations dealing with alert overload purchase dedicated solutions to track alerts. These SIEM (Security Incident and Event Management) solutions are expensive and often require specialized, dedicated technical talent. A few SIEM vendors claim to have solutions that sift through millions of alerts to provide a very short list of critical items that need immediate attention.

Addressing massive alert volumes is the focus of attention and innovation in Cyber today as most Infosec organizations are only detecting a fraction of their anomalies. In other words, while Detection is critical to the success of Infosec, no one

---

[4]Alert Fatigue, also called *Alert Overload*, happens when an Infosec team gets more alerts than they can respond to, or in many cases, even review. Some alerts, which could be critical, are ignored.

has built a solution that reliably detects every anomaly and accurately prioritizes those requiring responses. Still, the fact that Detection solutions prioritize alerts allowing Response Teams to stop some attacks in their earliest stages, can make these solutions worthwhile despite the challenges.

Prevention

Until the early 2000s, Cyber thought leaders believed 100% prevention would eventually be possible. Today, we believe that prevention, while indispensable, may never be 100% effective.

Prevention is about stopping attackers before they penetrate defenses and do their evil deeds. The walls of a medieval castle are a good example: They stop attackers. A determined enemy can scale them or smash them with cannon fire, but they keep most attackers out. To stop the most determined attackers, castle defenders added multiple walls, moats, and other obstacles. Similarly, Cyber prevention solutions have several layers designed to stop determined attackers.

Despite some marketing claims, **Today's Prevention measures stop all attacks some of the time, and some attacks all the time, but they cannot stop all attacks all the time.** Ideally, Prevention stops enough attacks so that an organization's Detection and Response capabilities are not overwhelmed.

An accurate count or inventory is the most important part of Prevention. A few years ago, I worked with an organization that suffered a near extinction-level breach caused by a half-dozen servers without any Cybersecurity software. The organization had acquired a second, smaller company that owned the half-dozen unprotected servers. The acquiring organization's Infosec team was not notified about the added servers after the acquisition. Over time, these servers were infected with many types of very nasty malware, partly because they had no Prevention measures in place and partly because the acquiring organization was a very attractive target. One day, the malware from these servers successfully attacked the organization's infrastructure. The resultant shutdown nearly led the organization to bankruptcy. The Infosec team first learned the unprotected servers existed

when they analyzed the breach. Ironically, the users who relied
on these servers (and their local server administrators) knew they
were infected, and that malware would crash them several times
each month. This information reached the Infosec team *after* the
breach.

*If your organization relies on the individual responsibility strategy
some call, "one throat to choke," you will want to understand
who is tasked with keeping an accurate inventory of all IT assets.
This task is rarely owned by Infosec although its accuracy is vital
for effective security.*

Budgeting by Cyber Discipline

In most organizations, asking the Infosec leader to explain
what percentage of their budget goes towards each of the three
disciplines is not likely to result in a useful answer. For
many of them, it may be the first time they've heard the
terms. Organizations budget for Incident Response Teams, but
Prevention and Detection spend is usually budgeted by asset
type, not by discipline. For example, Prevention and Detection
solutions for workstations are part of the workstation security
budget. Prevention and Detection solutions for networks are part
of the network security budget. And so on. Cybersecurity is
a collection of silos, each with their own Prevention, Detection,
(and in some cases Response) budgets, teams, and capabilities.

Chapter 4 reviews Prevention, Detection, and Response for each IT asset silo. In short, a complete Infosec strategy requires implementation of the following:

- **Response** – Documented plans for dealing with every likely attack type, from credential theft to Ransomware. These plans should be tested to ensure they restore business operations with as little downtime as possible.
  Boss Line: *If it's been on the news, we should have a plan to deal with it when it happens to us.*

- **Detection** – Solutions in place that detect everything that gets past Prevention measures without overwhelming the ability to review every significant alert.

  Note: *This is an aspirational goal for organizations dealing with Alert Fatigue.*

- **Prevention** – Solutions in place that stop enough attacks to make effective Detection and Response workable. If Detection is finding an unmanageably large number of successful attacks, Prevention is not stopping enough; it needs bolstering.

Agility

Until recently, Infosec organizations used a *Crown Jewels* strategy that relies on the fact that not all data is equally valuable. Systems with the most valuable data – the CEO's laptop, the Legal team's server, for example – got the best Prevention and Detection solutions and the most comprehensive and fastest Response. Lower-value systems got little or no Infosec investment, particularly in Detection and Response.

Ransomware takes advantage of the *Crown Jewels* strategy. A Ransomware attack could prevent the accounts receivable department from working, possibly crippling the business. The workstations and servers used by AR clerks are not considered *Crown Jewels* and are therefore less protected.

Ransomware changing the game for Infosec illustrates two of the biggest challenges Cyber practitioners face: The changing threat landscape[5] and the lack of agility in Cybersecurity solutions. Attackers revived Ransomware and invalidated the *Crown Jewel* strategy because it targets operations, not the most valuable data. If Infosec was as agile as modern software development, it could adapt to this change in days or weeks. However, the dozens of disparate Cybersecurity solutions (from a multitude of vendors) seldom interoperate and lack automation. This lack of integration makes it difficult for Infosec organizations to adapt quickly to a changing world.

Recap

- Cybersecurity condenses into three Core Disciplines: Prevention, Detection, and Response.

- Response measures are the most business-critical investment because Prevention measures today cannot not stop all attacks.

- A plan to reduce or eliminate the impact of a successful attack is the heart of effective Response.

- Detection is an important investment because many data theft attempts cannot be detected without them.

- Early Detection matters. As a rule, the sooner an attack is detected, the lower the cost of the Response.

- Detection is a nascent Cyber Discipline that occasionally fails. Infosec organizations are working on ways to manage enormous – and usually incomplete – alert volumes.

- Prevention measures are required because they stop most attacks. Without any Prevention measures, Detection and Response would be overwhelmed.

---

[5]Threat Landscape describes the sum of attacks and attackers facing Infosec defenders.

4

Protecting IT Assets

Effective Infosec is about applying the Three Core Disciplines (Prevention, Detection, and Response) to each of the following Information Technology asset silos:

- Identity

- Workstations and Internet of Things (IOT) Devices

- Network

- Servers or Workloads (if applicable)

Large Infosec organizations are most often divided into groups dedicated to protecting specific asset silos. As you might guess from the list above, there are identity security teams, workstation and IOT teams, network teams, and server / workload teams. Each team specializes in their particular assets. They evaluate and manage Prevention, Detection, and Response solutions for their respective silos. Even if there is a single Incident Response team, that team will have members specialized in each asset type. While there are some individuals with expertise in multiple silos, most specialize in one or two. For example, a network security expert is probably not ideally qualified to address a complex identity security issue, or vice versa. In smaller organizations, generalists handle multiple silos, but they often need outside help. Cybersecurity is such an enormous knowledge domain that few humans have deep expertise in every asset type.

I will explain what each silo is and what attackers want with it. I will then review each of the Core Disciplines, as they apply to each asset type, starting with Response. As a business leader, you only need a conceptual understanding of each of the asset silos covered below and of the three Core Disciplines from Chapter 3. Armed with that knowledge, you can have a productive conversation with any Cybersecurity leader, as outlined in Chapter 6.

Identity

What is it?

Every organization has user accounts, also called user identities. Identities include usernames with one or more passwords. Today, most users have multiple work-related accounts, so the number of accounts is much higher than the employee count.

For example, a company might have some 6,800 user identities for about 1,600 users. Manually managing this many accounts is difficult and can lead to lapses in security that attackers exploit. It is best to rely on an automated ID management solution.

In this book, ID theft refers to user credential theft, where an attacker gets username or certificate information to circumvent Cybersecurity measures. Consumer identity theft, where criminals impersonate a consumer to exploit their creditworthiness, typically targets individuals, not organizations.

What do attackers want?

Attackers want to impersonate users because impersonation lets them bypass many security measures. And, if the user they impersonate has powerful access rights, it saves attackers the added work required to covertly upgrade access rights.

Impersonating a powerful user lets attackers steal data, disable security defenses, plant malware for future use, and anything else their criminal minds dream up.

Response

When user credentials are compromised, the options are to disable the user account or reduce their access rights. In most cases, organizations elect to disable the account.

Here's an example of an automated version of this process. I was a US employee for a London-based company with staff in the US, UK, and Germany. On a trip to Germany, I had a layover in Paris for a few hours. There, I used my laptop to access our HR system and I got an odd error when I tried to log in. Next thing I knew, I was locked out of all corporate systems. The company expected my logins to originate in the US, UK, or Germany. When I logged in from France, the system assumed my identity had been compromised and disabled all my accounts. I was able to restore my access (and add France to the list of places I could work from) with a phone call to IT. Of course, they asked me several questions to ensure it was really me.

Detection

There are simple and complex Detection measures for User Identity Security. Simple detection involves notification and delay. For example, changing the phone number associated with an account triggers an email notification to the user and does not go into effect for 24 hours. This way, if the change is fraudulent, there's a good chance the legitimate user will flag the change as invalid.

Years ago, if a user tried to log in a second time (from a different system) using the same user ID, it was assumed that one of the users was a nefarious impersonator. This method is still used on some very sensitive systems, but it is not practical for most applications. Users today have multiple devices, all logged in and simultaneously accessing the same data.

Complex Detection measures for User ID security involve solutions from vendors specializing in Identity Access Management (IAM), Identity Governance & Access (IGA),

Privilege Access Management (PAM), or the more esoteric User and Entity Behavior Analytics (UEBA). Entire books have been written on each of these solutions. The basics are summarized in Chapter 5.

Prevention

As with every asset class, Prevention for identities begins with a complete inventory of every login for every user. This list should include all internal systems as well as any Internet-based (cloud) systems. There are software solutions that can automate this substantial administrative task, but it is still done manually in many organizations.

The most common user ID Prevention measures are frequent password changes and multi-factor authentication or MFA. MFA includes two- and three-factor authentication, abbreviated 2FA and 3FA, respectively. As the names imply, 2FA and 3FA add a second and third password-type credential required for successful authorization or login. These added credentials include short-lived, automatically generated passwords texted to phones or emailed to assigned user accounts. They can also include biometric information like fingerprints, voice and facial recognition, or retina scans.

The most basic ID security is something you know – a password. 2FA adds a second password, often automatically generated, and sent to something you have – typically your phone or email account. 3FA can add a third item used in lieu of a password, like facial recognition. To recap, conventional passwords usually rely on *something you know* (that criminals could learn). In the example above, 2FA adds *something you have* (that criminals could steal or clone) and 3FA adds *something you are* (which can be difficult to clone or bypass). 3FA provides stronger ID theft deterrence than 2FA, but neither provides 100% prevention against credential theft.

Passwords without 2FA/3FA should expire quarterly and not allow password re-use. It is best to avoid names or dates that an attacker can find. In 2018, several celebrities had private pictures published by attackers who guessed their passwords using public information (nicknames, pet names, birthdays, etc.).

Many organizations provide their users with 2FA as an option to reduce the chances of their identity being compromised. Cybercriminals are using social engineering[6] to add their phone numbers to the accounts that have not configured and enabled 2FA. If these criminals also obtain the passwords to these accounts, they can hijack them and shut out the legitimate users. The legitimate users then have the difficult task of convincing the organization that they are victims, not criminals. In short, if you are offered 2FA by an organization with access you value, enable it before a cybercriminal does.

Password length matters, especially if you don't have 2FA or 3FA. Eight or ten characters is an ideal minimum length. Shorter passwords can be guessed or broken by humans or computers. Much longer passwords are forgotten (which is bad) or written down (which is far worse). Using a browser to store passwords is relatively secure; using a dedicated password manager is better.

Don't make lists of your accounts and passwords and save them in a file. So many people do this (typically in Excel) that cybercriminals have malware designed to steal these files. Also know that password-protecting these files will only stop the least sophisticated attackers from reading them.

Workstations and IOT Devices

What are they?

Workstations are the devices and systems where users access and create data. Today, there are two main types of workstations: Conventional and smartphones / tablets. Conventional workstations are most often Windows or Macintosh computers with local storage and full keyboards. The line between tablets and conventional workstations is somewhat blurred today; tablets have more in common with smartphones than with workstations. Tablets and smartphones are designed primarily to allow users to display and interact with data. Conventional workstations are designed to allow users to **create**, display, and interact with

---

[6]Social Engineering involves using psychological manipulation and "con artist" trickery to get an individual to divulge information or to act on behalf of an attacker.

data. You may be reading this book on a tablet or smartphone, but I created it on a conventional workstation.

The Internet of Things is not a separate Internet for devices. Rather, it is a term used to describe the growing number of Internet-connected smart devices. These devices include webcams, drones, printers, TVs, monitors, thermostats, and a whole host of industrial, medical, and home automation equipment. While a bit odd grammatically, these network-connected devices are called *IOT devices*. IOT devices are the fastest-growing category of Internet-connected devices and attacker favorites because they are often unsecured.

What do attackers want?

Attackers are after data on workstations or a beachhead to launch attacks. Workstations and IOT devices have both been used as beachheads. Workstations provide attackers with several options. By clicking on a malicious web address (URL[7]) or by opening a compromised attachment, a user can unknowingly infect their workstation with malware. This malware can ransom or steal workstation data or establish a beachhead to launch attacks on other assets.

IOT devices don't have users clicking on links or email attachments, but they are usually very easy targets. Many organizations don't bother changing the default passwords for some IOT devices. These passwords are known to cybercriminals.

Response

For workstations, there are three mainstream options:

- **Rebuild** - Reinstall the operating system and applications from original media or a known-good image. Reinstalling virtually guarantees a clean system. The downsides to this option are that it can be time-consuming, and it deletes all local user files and customization.

- **Restore** - Restore the system from a backup. This option is usually faster than a full rebuild. The downsides include

---

[7]Uniform Resource Locator - A set of technologies used to find things on the Internet. It includes web addresses like http://www.fieldstargroup.com

not knowing if the backup was compromised, and loss of data (files) created since the backup.

Boss Line: *If we were recovering from a Ransomware attack, how quickly could we restore all of our systems?*

- **Remove the malware** - This is the fastest option, but there are no guarantees that all malware components will be removed. Sophisticated malware has components designed to restore it when most other components are deleted.

IOT devices usually cannot be backed-up, and malware removal tools don't work on most of them. Rebuilding is often the only option. It is vital to make sure software for rebuilding your IOT devices is available. I worked with an organization that had to replace nearly 100 ceiling-mounted webcams because they were compromised and there was no practical way to reinstall the software.

Detection

For conventional workstations (laptops and desktops), Detection can take place on the device itself or be deduced from network traffic. For on-device detection, the standard solution today is Endpoint Detection and Response (EDR). EDR detects workstation attacks that get past Prevention solutions. Virtually every endpoint security vendor sells a matching EDR solution.

For smartphones, tablets, and IOT devices, the most popular Detection option is to deduce attacks based on their communications. Solutions that scour network traffic to find attackers' communications can detect compromised devices. These solutions, now called Extended Detection and Response or XDR[8], monitor network traffic for signs of compromised desktops, laptops, servers, smartphones, tablets, and IOT devices. They are not as effective as EDR, but they represent the state of the art in Detection for devices where EDR is not an option.

---

[8]XDR is a relatively new acronym. Some vendors used Network Detection and Response (NDR) or Network Traffic Analytics (NTA) before the industry settled on XDR.

Prevention

Prevention for workstations and IOT devices begins with an accurate inventory. Every device used to access your data and network needs to be counted. Device inventory info is required to effectively manage the four available Prevention options: Network, Device, User, and Locally Installed.

Network Prevention is about using network devices (like firewalls) and network designs to keep network-connected devices safe while allowing them the network access needed for their intended purposes. This option is only available for organizations with private networks.

Network isolation, also called segmentation, can effectively segregate traffic providing improved workstation and IOT device security for organizations with large in-house networks. The details are described in the Network section below.

Device Prevention is about ensuring that any device handling your data is authorized to do so. No unauthorized device should ever access your data, or at least your *most sensitive* data.

Devices are authorized using certificates, specialized software *permission slips*. Accessing the network or an application can require a valid certificate installed on the workstation or IOT device. For example, accessing the organization's private network could require a certificate, and accessing email could require a second. In most cases, certificates are used along with user IDs and passwords. Requiring both means the device requires a valid certificate and the user on that device would also require username and password(s). Using both provides better Prevention and adds administrative overhead.

In theory, certificates prevent any unauthorized device from accessing specific assets. Certificates can be configured to work only on one specific device; copying them to a second device would not give that second device access. In practice, certificates are often configured so they will work in many similar machines (laptops remote users own, for example) to reduce administrative overhead. If this is the case, an attacker with an unauthorized device only needs a copy of a certificate. Also, while a valid certificate confirms the machine is authorized, it does not ensure that it is not compromised. As an example, imagine a laptop with a valid certificate that allows it to connect to sensitive servers. That laptop connects to a public Wi-Fi network and gets infected with Ransomware. When the laptop returns to the office, the certificate still allows it to access the sensitive servers. The Ransomware also has access; it could encrypt the data on the sensitive servers and demand ransom.

*Certificate management solutions for devices are called **Machine Identity Management** solutions.*

User Prevention is about ensuring that any individual viewing or handling your data is authorized to do so. No unauthorized individual should ever view or access your data, or at least your *most sensitive* data.

For workstations, this subject is covered in detail in the Identity section above. For IOT devices, User Prevention measures involve changing the default username and password – attackers know these.

Locally Installed Prevention solutions, as the name implies, are installed on the devices themselves. These solutions can be highly effective, but they consume resources reducing the capabilities and performance of the systems they protect.

Locally installed Prevention for conventional workstations is the oldest type of Cybersecurity. Endpoint Protection Platforms (EPP) from vendors like Symantec, McAfee, Microsoft, Trend Micro and dozens of others are common. These solutions evolved from, and still include, Anti-Virus (AV) software. Many security practitioners refer to these solutions as AV because of their origins. None are 100% effective, but they stop many attacks.

There are EPP-like on-device solutions for smartphones and tablets, but most organizations don't use them, or use them only for their most critical devices (like the CEO's smartphone). These organizations rely on network isolation (Network section below), certificates (Device Prevention section above), or Detection and Response measures to contain attacks involving these devices. There are known attacks that steal data from smartphones and tablets or use them to launch attacks on other assets. Today, the consensus in Infosec has most organizations not providing on-device Prevention for these devices. There are solutions, but there is little innovation in the space. Some available solutions impact device performance and hurt battery life.

A future well-publicized attack originating from smartphones or tablets could dramatically increase demand for on-device Cybersecurity solutions for these workstations – and would likely drive innovation and product improvements.

For most IOT devices, on-device Prevention solutions are not an option; these devices lack the power and features required to support added software solutions.

Locally Installed Patches

Workstations and IOT devices have vulnerabilities that can be exploited by attackers. A vulnerability is a software flaw that allows an attacker to disable or bypass security measures. When vulnerabilities are made public, the vendor responsible for the software issues a software update called a *patch*. Applying the patch updates the software and removes the vulnerability. Let's look at an extreme example from a few years ago.

Experts believe that around 2014, the US National Security Agency (NSA) discovered a vulnerability in Microsoft Windows. The vulnerability existed in every version of Windows at that time and allowed remote attackers to take administrative control of any Windows server or workstation without anyone noticing. The NSA built a tool to use this vulnerability and named it Eternal Blue. The NSA (stealthily) used Eternal Blue for about three years. There are also reports that the NSA shared Eternal Blue with allies – no word on what *they* did with it. In March 2017, attackers published information stolen from the NSA, including Eternal Blue. Within days, Microsoft issued Windows patches removing the vulnerability, thereby neutralizing Eternal Blue.

If you think that was the end of the story, you'd be wrong. In May 2017, nearly 60 days after Microsoft's patch rendered Eternal Blue harmless, millions of Windows systems all over the world were hit with WannaCry Ransomware. WannaCry uses Eternal Blue to bypass Windows defenses. In June and

September 2017, two other attacks based on Eternal Blue hit Windows systems (servers and workstations) around the world. Their impact was less than WannaCry, but still significant. The attackers behind WannaCry did not get the information from the NSA, or from the thieves that stole Eternal Blue. Instead, they got it from the documents Microsoft provided with the patch. Attackers study Microsoft's patch information to develop new attacks, because organizations are slow to patch their systems.

Workstation patch installations are delayed for two primary reasons: scale and prior pain. There are dozens of patches each month, from OS vendors and application vendors. Deploying them to large-scale organizations is a huge task. Also, practitioners have deployed patches that led to (sometimes massive) system failures and outages. As a result, patches are usually installed on a small subset of workstations first and allowed to run for a few weeks before deploying to the overall population.

For attackers, this means there is window between the release of a patch and its widespread implementation; a window where they can write and execute new, and highly effective, attacks.

Many IOT devices do not support patching; their vulnerabilities are permanent.

Network

What is it?

The Internet is a global public network connecting millions of private networks. It is unlikely that any device you use today connects directly to the Internet. Your smartphone connects to your cell phone provider's private network, which connects to the Internet. Your home's broadband router connects to your Internet Service Provider's (ISP) private network, which connects to the Internet, etc.

Originally, organizations licensed applications like email, finance, and HR and hosted them on in-house servers. These servers were connected to each other and to workstations using private networks. Specialized boxes connected all the wires and wireless signals. These private networks then connected to the Internet through one or more Internet Service Providers' private networks. Firewalls were installed between private networks and ISP networks (and the Internet) to keep cybercriminals out.

Today, Cloud-based Software as a Service (SaaS) applications like Salesforce.com, Oracle Cloud, and Microsoft Outlook, make it possible for organizations to function without in-house networks. If your organization has an in-house, private network, there are Response, Prevention, and Detection measures covered below that will improve your security posture. If, on the other hand, your staff works entirely from home, and you rely on SaaS solutions, your security measures need to focus on the other asset silos. Home users have their own small, private networks, but securing all of them is not practical.

What do attackers want?

While there are some rare, specialized attacks targeting private network components, most cybercriminals use networks the same way legitimate users do: To move data and remotely control systems. In Cybersecurity's early years, network security experts thought their firewalls were very effective and that few attackers got past them. This led to a security model called *Trusted Networking*. The private network (inside the firewall) was *Trusted*; everything outside the firewall (e.g., The Internet) was not trusted because cybercriminals lived there. Belief in Trusted Networks had many organizations doing things we consider irresponsible today. For example, most servers in a Trusted Network had little or no security running on them: No Prevention, no Detection. Attackers that made it past the firewall – and there were many – had a field day. Even today, although Trusted Networks are passé, an attack originating inside a private network has a much higher chance of succeeding than the same attack originating from a system on the Internet. This is why attackers compromise systems inside organizational firewalls. The attacks they launch from *insider* systems are much more likely to succeed.

Response

There are two primary network-based responses to attacks: Exfiltration blocking and Isolation. Exfiltration is the term used to describe the act of data theft. When the cybercriminals are moving your data through your firewall to their servers on the Internet, that's exfiltration in progress. Advanced (Next Gen) firewalls have exfiltration detection and can block it. As a business leader, if your network has valuable data, you'll want to know you have solutions in place that can detect or stop exfiltration.

Isolation, as the name implies, limits or removes network access from a suspect device. I had an engineer who downloaded a hacking tool to see how it worked. The tool was designed to find network devices with poor security. He ran it on our corporate network. The Infosec team had an automated solution that detected the tool's activity almost immediately and isolated his system. The PC with the hacking tool suddenly couldn't access anything on the network. This is a good example of effective, automated network isolation because the detection solution assumed the engineer's PC was launching an attack.

Detection

Network-based detection measures include the following:

- **Connection Detection** solutions track all connected systems and (often manually) flag unauthorized systems.

- **Machine ID** solutions can notify when unauthorized systems connect to the network, but they are usually configured to keep unauthorized systems from connecting.

- **Exfiltration** solutions, most often included in Next Gen Firewalls, can notify Infosec when sensitive data is leaving the private network.

- **eXtended Detection and Response, XDR** solutions monitor network traffic looking for compromised systems

- **Intrusion Detection** systems look for evidence of attackers trying to breach network defenses (e.g., firewalls).

Prevention

Accurate and up-to-date network drawings provide the key starting point for network Prevention. Frequent changes and company acquisitions not added to network plans have led to some of the costliest breaches. From an inventory standpoint, a listing of every network is the minimum requirement, a complete list of every network-connected device is ideal.

Beyond accurate and complete inventory, network Prevention focuses on two questions:

- Who and What can connect to our network?

- What assets can they then access?

Network Connection Prevention

With wired networks, connecting is about plugging a cable into a wall jack. There are solutions that require verification before granting access once a system plugs in, but they are rarely used. The fact that wired network jacks are in buildings with some degree of physical security is sufficient for most Infosec teams.

Wireless (Wi-Fi) access should be encrypted and require long, complex passwords. Some experts recommend changing Wi-Fi passwords regularly, which can be disruptive in many organizations. Others recommend making the Wi-Fi password a secret known only to Infosec or IT. This is only practical in organizations where individual users are not the Administrator or Super User for their workstations – which can lead to help desk issues.

Getting rid of, or disabling, firewalls allows remote users full access to all assets as if they were in the office. Formerly internal assets would be attacked in seconds by thousands of cybercriminals. Instead, we give remote users specialized software that creates a VPN or Virtual Private Network. VPN software lets users connect into private networks without exposing the private network to the Internet. VPN software does nothing to ensure the integrity of the system. It will allow communications between a remote workstation and the private network even if the workstation is launching a sophisticated attack.

Network Access Prevention

Beyond network access, we get into the somewhat arcane world of network design. Know that determining which assets a user can

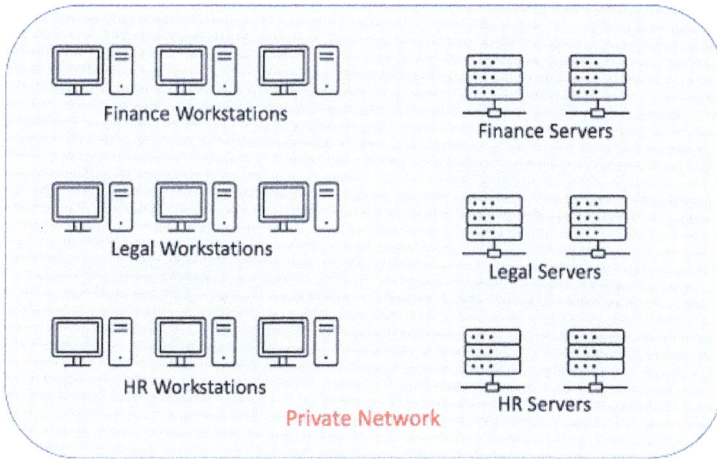

Figure 4.1: *Private Network*

access once they're connected to a private network is complicated stuff. If you're interested, the next paragraphs are for you. If not, skip ahead to the **Servers or Workloads Section.**

After connecting, network design determines Prevention. Let's look at an example in Figure 4.1. We have three groups of users in our private network: Finance, Legal, and HR. Each group has some PCs – where users work – and some servers where the group's data is stored. Note the network wiring is not shown. Assume every device can communicate with every other device unless they are separated by one or more lines.

Without any network security, any user can access any server inside our private network. A user from HR can access the Finance server – even if they have no business there. That HR user would need credentials (login, password, and possibly a certificate) to access the data on the Finance server, but they could access the server on the network giving them (or malware on their system) the opportunity to hack into it.

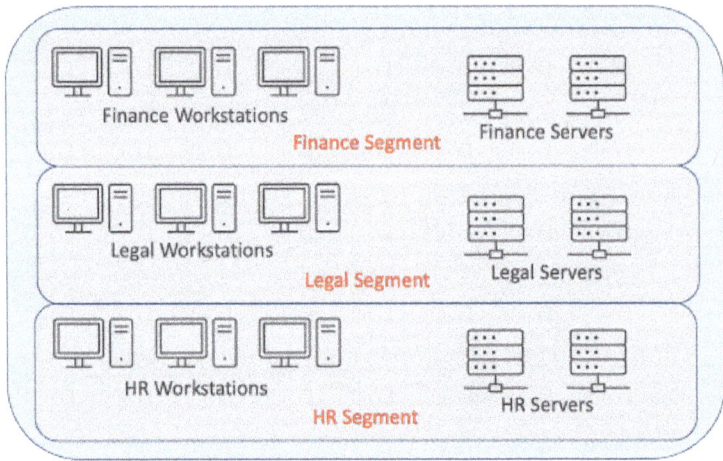

Figure 4.2: *Segmented Private Network*

The most basic network security is called segmentation or isolation. As you can see in Figure 4.2, each of our three user groups is isolated in their own network segment. A user from an HR workstation can no longer access the Finance server. For IOT Devices, network isolation is often the only available Prevention measure. For example, the webcams in the ceiling of a retail store can be isolated so they can only communicate with other webcams and the server that records all the video streams. This way, if a webcam is compromised by an attacker, it can only infect other webcams or the video recording server. Similarly, smartphones and tablets can also be placed in their own segment. If these phones and tablets need access to in-house network-connected assets, granting that access reduces (or possibly eliminates) isolation. For example, letting smartphones and tablets access the Internet from their own isolated segment would keep a compromised smartphone from attacking any corporate asset excepting other smartphones and tablets. If, however, the smartphones also needed access to the in-house HR server, that server would be vulnerable to an attack from a compromised phone or tablet.

Now let's assume users work from home and connect to the Internet. If we opened our servers to the Internet, they would be compromised in seconds. Instead, we give our remote users

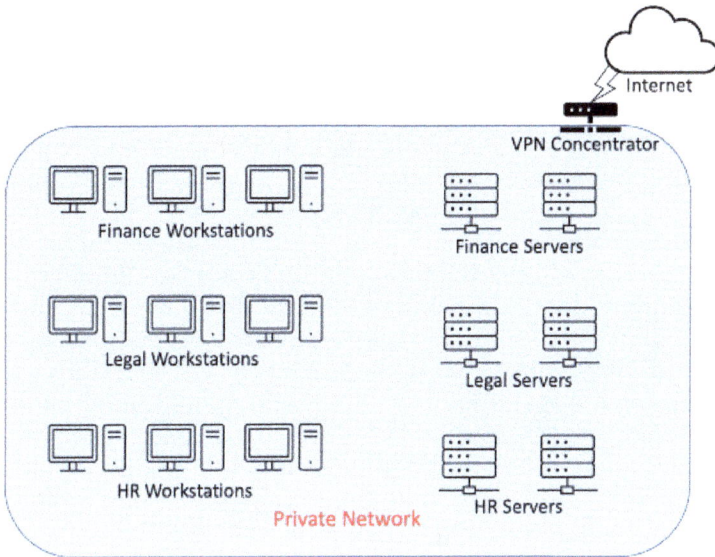

Figure 4.3: *Private Network with VPN*

specialized VPN software. This keeps our servers safe from attackers on the Internet while allowing our users to access them.

In Figure 4.3, you may have noticed the segmentation from Figure 4.2 is gone. The single VPN connection point, called a VPN concentrator, needs network access to every server because remote Finance, HR, and Legal users (in this example) all connect to the private network using this single concentrator. Every user can once again access every server. Buying, installing, and maintaining three VPN concentrators (one for each segment in Figure 4.2), preserves segmentation. But larger organizations have hundreds or thousands of segments, making VPN concentrators for each segment cost-prohibitive.

VPN software keeps Internet attackers from reading or hijacking our remote users' connections to our network and servers, but it does nothing to prevent the remote users' systems from being compromised.

I worked with a Fortune 500 company that spent decades refining their network segmentation. Nearly all their users worked at corporate sites. Most users could access the company email system, the Internet, and the in-house servers they needed for their jobs. All other access was blocked at the network level. This segmentation – or isolation – limited access and prevented a compromised individual or system from attacking valuable targets. In short, it was state of the art network segmentation security. In the 2020 pandemic, management sent thousands to work from home for the first time. The IT department scrambled to buy, install, and configure VPN concentrators for the entire workforce. They did not have time or the budget to provide dedicated VPNs for each group – which would have preserved their highly secure isolation. Instead, every VPN-connected user could access nearly every system on their network. Days into the new VPN implementation, the Infosec team detected dozens of attacks on their most valuable servers. The attacks originated from home users with compromised systems. The company had ordered laptops with advanced security, but deliveries were delayed for months. Managers allowed remote users to install the VPN software on their own (often compromised) systems so they could work while they waited for the new laptops. The attacks did not result in data loss or disruption, but they overloaded the already stressed, newly remote Infosec team for months.

Experts believe that segmentation and VPNs will ultimately be replaced by Zero Trust solutions. In Zero Trust networking, access to any asset is only granted to users or systems that specifically require that access–regardless of location. As the name implies, Zero Trust does away with the concept of Trust — the design principle in Figure 4.1. In a trusted network, all of the systems are trusted and therefore have network access to all other systems. The fact that attackers have little trouble getting into private networks obsoleted the Trust model years ago.

Servers or Workloads

What are they?

Not long ago, organizations had many of their own in-house servers hosting email, applications, web servers, file sharing, and many other things. Today, most of those things are available as services in the cloud, making it possible for organizations to operate without servers.

Originally, a server was a physical computer box (essentially an oversized PC) hosting a single application shared by many users, perhaps an email server. Starting about 15 years ago, specialized software allowed that oversized PC to host multiple servers, with each server isolated from the others. These servers were called *virtual servers* to differentiate them from servers running on dedicated boxes. In the early days of virtualization, the term *workload* was used to differentiate between virtual servers and legacy servers. Today, the overwhelming majority of servers are virtual servers, and the names *server* and *workload* are used interchangeably.

What do attackers want?

Servers are where in-house data is stored. If you have an in-house CRM (Customer Relationship Management) solution, your CRM servers are where your customer data lives. If you have an in-house email server, all emails are stored there. And so on. Attackers are interested in servers because of the data they contain, just as bank robbers are interested in money.

Response

Like workstations, workload response has the same three options. Workloads have the added challenge of supporting multiple users. A compromised server may still work partially, providing users with some functionality. These Responses, except for removing the malware (which is not recommended) will take the server offline for a time.

- **Rebuild** - Reinstall the operating system and applications from original media or a known-good image. Reinstalling virtually guarantees a clean system. The downsides to this option are that it can be time-consuming, and it still requires restoring data from backup.

  Restore - Restore the system from a backup. This option is usually faster than a full rebuild. The downsides include not knowing if the backup was compromised, and loss of data (files) created since the backup.
  Boss Line: *If we were recovering from a Ransomware attack, how quickly could we restore all our workloads?*

- **Remove the malware** - This is the fastest option, but there are no guaranties that all malware components will be removed. Sophisticated malware has components designed to restore it when most other components are deleted.

Detection

As with workstations, EDR is an option for workloads. Many prevention solutions (workload- and workstation-specific) provide Detections. Adding XDR to EDR provides an added layer of Detection.

Worstation EPP solutions include anti-virus and several other Prevention and Detection modules, workload-specific solutions have the following core modules:

- **File Integrity Monitoring or FIM** monitors critical files to ensure they are not compromised by attackers.

- **Configuration Security Monitoring or CSM** reports any configuration changes that could be signs of a future or ongoing attack.

- **Vulnerability Management** Reviews available patches. Some solutions also include **Patch Management** modules.

Prevention

Like with workstations, accurate workload inventory is crucial. This is complicated in environments that use rapid in-house application development. Today's development techniques quickly respond to changes in user counts and functionality by creating and deleting workloads as needed. As a result, workloads may only exist for hours or days. Also, there is a new component called a *Container*, which is essentially a mini-workload. Containers can be created or deleted in seconds and often only run for minutes. Containers also have the added challenge of requiring container-specific security solutions. Workload security software will not effectively protect them.

In organizations without in-house application development protecting workloads is about selecting one of two options: Treat the workloads like workstations and use the same Prevention measures or use dedicated workload security solutions. For workloads and servers there are three available Prevention options: Network, Device, and Locally Installed.

Network Prevention is about using network firewalls and network designs to keep network-connected devices safe while allowing them the network access needed for their intended purposes. These measures were described in the Network section.

Device Prevention is about ensuring that any device handling your data is authorized to do so. No unauthorized device should ever access your data, or at least your *most sensitive* data.

Instead of user accounts, workloads rely on certificates to authenticate on (e.g., log in to) other servers. Certificates are specialized software permission slips, as explained in the **Workstations and IOT Devices – Prevention** section in Chapter 4. Certificates can also encrypt communications between servers, making it very difficult for attackers to eavesdrop on a conversation. Misuse of this Prevention measure may have led to one of the most expensive breaches in history.

To block the exfiltration of valuable data (explained in the **Network – Response** section in Chapter 4), security solutions designed to stop data from being stolen must be able to read it. This presents a problem when server communications are encrypted. Encryption prevents anyone without the password key from reading the content. If the exfiltration-detecting device cannot read the content of the communications, it cannot detect exfiltration. To resolve this, we can install a matching certificate (matching the certificates on the servers) on the detecting device. This certificate provides the detecting device a key to read the encrypted communications.

Certificates expire; they need to be replaced or refreshed every 12 months or less. Many organizations have manual or incomplete certificate management, and they often use servers with expired certificates. Applications can ignore the expiries with the working assumption that the certificates will eventually be updated.

Expired certificates (allegedly) had a big part of one of the most expensive breaches in history. One story says the organization that suffered the major breach had a server in their private network used for low-priority communications with a server in another organization. Communication between the two servers takes place over the Internet and could therefore be intercepted – so, it was encrypted. Each server had a certificate making it so. To ensure no sensitive data was exfiltrated, the organization used a Next Generation (NG) firewall with a copy of the certificate. This allowed the firewall to read the data exchanged by the two servers – and detect exfiltration.

A few months before the breach, the certificates on the two servers expired, and the organization did not replace or update them. The servers were still able to communicate with each other, and that communication was still be encrypted. However, the NG firewall will not work with expired certificates. It blocked communication between the servers because it could no longer inspect the traffic for sensitive data exfiltration.

When the firewall stopped the two servers from communicating, the line of business managers responsible for the servers demanded communication be restored. The network security team leader explained that their only option was to reconfigure the firewall allowing encrypted communications using expired certificates to pass through the firewall uninspected. Notice that in this story, nobody contacted the team responsible for certificate management. That team could have updated the certificates for the servers and the firewall, resolving the problem without impacting security.

The LoB teams' request led to a change in firewall exfiltration policy. Going forward, communications using expired certificates would pass through the NG firewall uninspected. You might have already guessed what happened next. Cyber criminals used an expired certificate to exfiltrate hundreds of megabytes of sensitive information. The sensitive data they stole went through a very expensive exfiltration detection solution – completely undetected. Ironically, the most expensive certificate management solution costs less than 5% of the hard costs of the breach. The management dynamic from this story exists in many organizations. Managers outside Cyber sometimes see Infosec as an unnecessary obstacle, and some Infosec managers knowingly weaken security measures to avoid political confrontations. The resulting breaches are avoidable and sometimes very expensive.

Certificates have three downsides:

- **Complexity and Cost:** Effective certificate management requires dedicated solutions and expertise.

- **Added Cyber Risk:** Improperly managed certificate management often creates exposure windows. One of these windows led to a breach costing over $1B (in hard dollars).

- **Over-reliance:** Effective certificate management enhances Prevention by reducing the possibility of unauthorized devices accessing the network or applications. It does not prevent those devices from being compromised. For example, an organization could require certificates on all smartphones accessing their network. This would prevent any unauthorized smartphone from accessing the network, but it would not prevent an authorized, compromised phone from carrying out an attack.

In short, Machine Identity Management is usually a requirement for organizations that own servers. If your organization relies on certificates for machine identity, make sure you have a well-designed (ideally, externally audited) Machine Identity Management process. Shoddy Machine ID management has contributed to some of the most expensive breaches in history.

Machine Identity Management solutions rely on certificates and Public Key Infrastructure or PKI.

Locally Installed solutions, as the name implies, are installed on the workloads. These solutions can be highly effective, but they consume resources, reducing the capabilities and performance of the servers they protect.

If all workloads run Windows, treating them like workstations has advantages in costs and administrative overhead: One solution for all workloads and workstations. In Linux or mixed Windows and Linux workload environments, dedicated workload-specific solutions may be a better option. Workstation solutions include Anti-Virus functionality (as explained in the Workstations and IOT Devices section) while most dedicated server solutions do not. In theory, workloads are less likely to be infected by viruses because there is no user to click on malicious links or email attachments.

Locally Installed Patches

Like Workstations and IOT devices, workloads have vulnerabilities that can be exploited by attackers. A vulnerability is a software flaw that allows an attacker to disable or bypass security measures. When vulnerabilities are made public, the vendor responsible for the software issues a software update called a patch. Applying the patch updates the software and removes the vulnerability.

Server Patches are delayed for three primary reasons: downtime, prior pain, and scale. Patches often require server reboots – which on some systems can take hours. These patches must be scheduled for off hours – with Sunday mornings before 6 am being a favorite. With servers running custom or older software, patches can disable applications or introduce strange bugs. As a result, practitioners are careful to only patch one system at a time, allowing it to run for a week or two (if possible). Some patches don't include removal instructions, and these must be created so there is a recovery option if the patch creates problems. These issues, combined with the scale of some organizations, makes workload patching a very slow process – often slower than workstation patching. Keep in mind that if the patch removes a significant vulnerability, measures should be put in place to prevent this vulnerability from being exploited – or to at least Detect if it has – until the patch is fully deployed.

Development Workloads

Workload or server security for organizations with in-house development is a complex topic. Know that while software development can now accomplish in days what took months years ago, Cybersecurity is not nearly as fast or agile. Because of this difference, developers tend to ignore Infosec until their projects are completed. At that point, they ask the Infosec team to bless their new application and complain when that blessing is not immediate. There are some Cyber solutions that feature agility and speed but getting application development to consider Infosec a partner – and not an unneeded end-of-project delay – is an important (and rare) first step.

Public-facing Web Servers

Like submarines – boats that sink on purpose – public-facing web servers are purposely exposed to the Internet in the name of E-commerce or customer service. Securing these specialized servers requires all the measures mentioned in the **Servers or Workloads** section and more. As a businessperson, be aware that these servers increase the risk of successful attacks. Public-facing servers often access sensitive data (customer info, credit cards, etc.) and are more likely to be compromised because of the increased **Attack Surface**[9] required to make these servers accessible to the Internet public. Chapter 6 includes some useful questions and talking points about these servers.

---

[9]Attack Surface is the name given to all the available options attackers have to compromise target systems.

5

Solutions Primer

This chapter covers the most used Cybersecurity solutions. As a business leader, you don't need to know all these details; they are provided as reference in case you're curious about specific expenditures.

Effective Cyber defense is about *Three Ps: People, Products, and Processes*. The following is a short primer covering commonly used Products and Processes. The brunt of vendor-supplied Cybersecurity products today support Prevention and Detection disciplines. Response is a set of procedures that leverages software solutions, most of which are not Cybersecurity-specific. For each asset category, I've included the most popular solutions, starting with Response. Prevention and Detection solutions are divided into *Basic, Advanced,* and *Cutting Edge* categories. The categories are my opinion; there are no defined standards. In the following figures (5.1 - 5.5), solutions in square brackets, like "[Patching]," rely mostly on procedures while the others rely mostly on specialized products.

Response

Detection solutions generate alerts that trigger Responses for each asset silo. The list below summarizes the most common responses.

Identity

If an Identity (user login) is found to be compromised, Infosec practitioners use one of the following options:

- Reduce user access rights for the compromised account

- Disable the compromised account (default)

Workstations and IOT Devices

If workstations or IOT devices are compromised, they must be restored to their pre-infection states. The three options below are the most common. Note: *Most IOT devices only support Rebuilds.*

- Rebuild from original media or image

- Restore from backup

- Remove malware (limited use only)

Network

If any network-connected device is found to be compromised, it should be isolated (curtailed communications) or disconnected from the network. Note that devices can be disconnected logically; there's usually no need to disconnect a physical cable. Some network solutions also detect exfiltration and can stop valuable data from leaving a private network. If stopping exfiltration is not an available option, disconnecting or isolating the compromised device can accomplish the same goal. In short, the Network Response options are:

- Isolate or disconnect from network

- Prevent Exfiltration

Compromised workloads, workstations, and IOT devices will need further response, as detailed in the **Workstation and IOT** *and* **Servers or Workloads** *sections.*

Servers or Workloads

Like with compromised workstations, compromised workloads must be restored to their pre-infection states. These are three most widely used options:

- Rebuild from original media or image

- Restore from backup

- Remove malware (limited use only)

For more information, refer to the Servers or Workloads section in Chapter 4.

Detection and Prevention

Identity Security

User ID Prevention measures begin with an accurate and up-to-date tally of all user accounts. As shown in Figure 5.1, Basic prevention includes quarterly password expiry, disallowed password re-use, and minimum password lengths. Advanced Prevention measures add 2 Factor Authentication, Identity Access Management (IAM) and Identity Governance and Access (IGA) solutions. Cutting Edge Prevention adds 3FA using Biometrics.

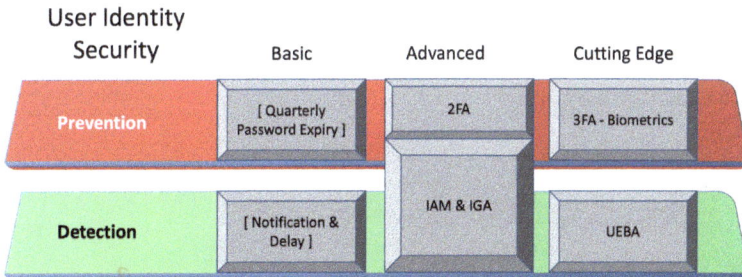

Figure 5.1: *User ID Security Solutions*

User ID Detection measures begin with *Notifications and Delay* policies. These policies notify the user about changes that could be triggered by attackers. If, for example, a phone number is added as part of a user's identity, that user should be notified so they can confirm the number is legitimate. Delay is also important, as we need to give the legitimate user time to confirm – or deny – the changes. Advanced Detection adds IAM and IGA, and Cutting-Edge detection adds User & Entity Behavior Analytics (UEBA). UEBA solutions track user behavior and alert Infosec when user behavior is unusual or suspicious. For example, if after 8 years the CFO has never logged in after 10 pm, having them log in at 3 am would generate a UEBA alert disabling her account and notifying Infosec.

Workstations and IOT Devices

Workstation measures (Figure 5.2) begin with an accurate and up-to-date tally of all devices that access the organization's data. Basic Prevention for laptops and desktops includes Patching and Endpoint Protection Platforms (EPP), commonly referred to as AV. There are Prevention solutions for smartphones and tablets, but they are not commonly used except for devices accessing the most sensitive data. Machine ID Management can provide added Prevention by ensuring only authorized devices are used. On the Cutting Edge of Prevention for workstations, there are new Neural Network solutions for PCs and laptops that appear to be more effective than traditional EPP solutions.

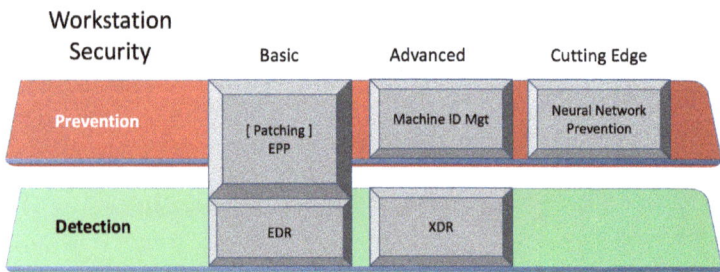

Figure 5.2: *Workstation Security Solutions*

Prevention for IOT devices (Figure 5.3) begins with a complete device inventory and requires changing the default usernames and passwords (as these are known to many attackers).

These devices should also be regularly patched if their manufacturer allows it. Added Prevention measures for IOT devices include network segmentation and Machine ID Management, as covered in Chapter 4.

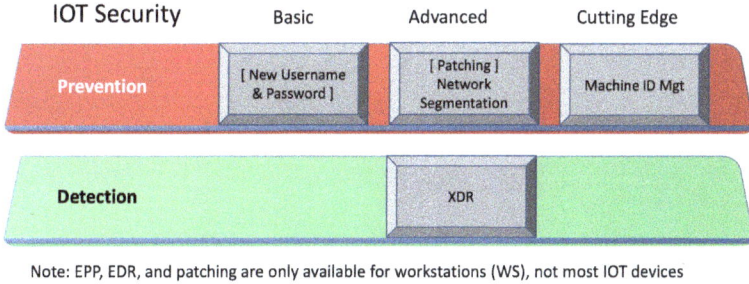

Note: EPP, EDR, and patching are only available for workstations (WS), not most IOT devices

Figure 5.3: *IOT Security Solutions*

Workstation Detection measures involve Endpoint Detection and Response (EDR) for workstations and laptops, and Extended Detection and Response (XDR) for all Workstations and IOT devices. EDR has proven highly effective at Detecting what EPP solutions miss, but it is limited to workstation PCs and laptops. XDR is a universal solution, supporting all network-connected devices, including smartphones, tablets, and IOT devices. Unfortunately, XDR solutions are expensive, complicated, and not as effective as EDR.

Network

Assuming an accurate accounting of network components, Network Prevention measures (Figure 5.4) begin with Firewalls (Basic) and currently ends with Zero Trust Networking (Cutting Edge). Smart or Next-Gen Firewalls represent the middle ground or Advanced Prevention. Any organization with a network needs at least a basic Firewall between the internal (private) network and the Internet. This device keeps out many attackers but should not be confused with the software firewalls included in virtually every workstation and server (as well as some smartphones and tablets). Those firewalls protect individual devices; network firewalls are designed to protect all the devices on a private network. A Next Gen (NG) firewall improves on

Network Protection and provides some sophisticated Detection. Zero Trust Networking – where every network connection is scrutinized – represents the future of network-based Prevention. Implementing Zero Trust in most large networks is expensive, difficult, and time intensive.

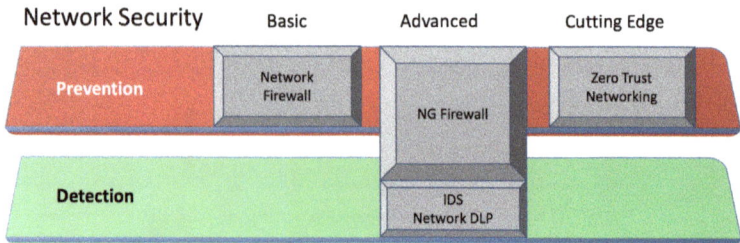

Figure 5.4: *Network Security Solutions*

Network Detection finds attacker behavior in the network as well as data exfiltration (sensitive data exiting the private network for the Internet). NG Firewalls and Intrusion Detection Systems (IDS) both alert on attacker behavior, NG firewalls also alert on exfiltration. Also, while rarely used in 2021, there are Network Data Leakage Prevention (DLP) solutions dedicated to detecting data exfiltration.

Servers or Workloads

As with workstation security, workload Prevention measures (Figure 5.5) begin with an accurate and up-to-date tally of all servers. Basic Prevention includes Patching and a dedicated security solution. That solution can mimic workstation Security using Endpoint Protection Platforms (EPP) or can be workload-specific. Advanced Prevention leverages virtualization to provide the same level of protection with significantly improved performance and reduced resource utilization.

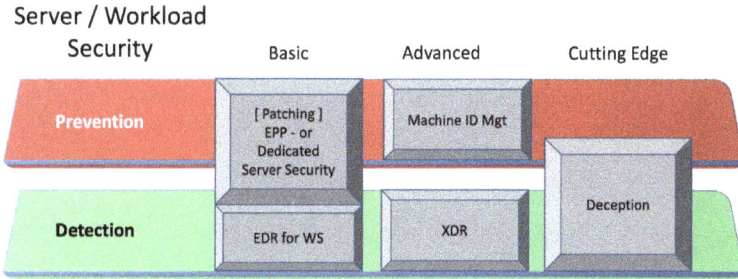

Figure 5.5: *Server or Workload Security Solutions*

Server-specific Prevention and Detection solutions include the following modules:

- **File Integrity Monitoring or FIM** monitors critical files to ensure they are not compromised by attackers.

- **Configuration Security Monitoring or CSM** reports any configuration changes that could be signs of a future or ongoing attack.

- **Vulnerability Management** Reviews available patches. Some solutions also include **Patch Management** modules.

Most workload-specific solutions do not include anti-virus functionality required for regulatory compliance on workloads with credit card information.

Workloads that communicate with other assets require certificates in lieu of usernames and passwords. For these servers, effective Machine Identity Management is a key Prevention component.

Again, as with workstations, workload Detection measures involve Endpoint Detection and Response (EDR) and/or Extended Detection and Response (XDR). EDR has proven highly effective at detecting what EPP and dedicated server solutions miss, but it requires software on each workload. XDR provides similar detection to EDR using network traffic, eliminating the need for software on each workload. Unfortunately, XDR solutions are complicated, and not as effective as EDR.

Deception technologies, which create fake workloads to lure attackers, represent the Cutting Edge of server Detection – and Prevention to a lesser extent. With a deception solution, we create multiple copies of a valuable target workload. Assuming the attackers only attack one server at a time, these copies make it statistically likely the attackers will ignore the real target. Also, the fake servers are highly instrumented; they record every detail of the attack. This information, if harvested quickly, can protect real targets from the same attacks.

Alert Management

While Alert Management is a bit outside the Core Disciplines discussed in this book, it represents one of the fastest growing - and most expensive - parts of Infosec today.

Organizations with large scale alert traffic invest in Security Incident and Event Management (or SIEM) solutions. For most organizations, a SIEM for each asset silo would be cost-prohibitive; the usual deployment is a single SIEM across all asset types. The SIEM can analyze historical data to find older (but possibly still active) breaches as well as new breaches. The SIEM should include *analytic* functions to prioritize alerts. Cutting-edge SIEMs use Artificial Intelligence (AI) and Machine Learning for prioritization and forensics. *Data Lakes* are also on the cutting edge. These free-form databases are designed to find attack clues in enormous sets of alert data.

6

Cyber Conversations

The following conversation outlines help business leaders understand the decisions made by Infosec leaders. The first conversation reviews the defenses in place, the second covers the main questions you should ask following a significant breach.

The business leader's talking points are italicized, while notes, comments, and observations use regular type.

Cyber Defense Review

Identity

I'd like to understand how we're protecting each of our organization's IT assets from Cyberattacks. Let's start with identities. I understand that if an attacker used my identity (or yours) they could bypass most of our security measures.

Identity - The Most Critical Questions

- **[Detection]** *How quickly would we detect a compromised user ID?*

- **[Response]** *What are our plans for that eventuality?*

- **[Response]** *Have we tested these plans?*

Identity Conversation

+ **[Risk]** *How much damage could an attacker using a compromised user ID do?*

+ **[Inventory]** *Do we have an accurate list of every login (including outside services)? Who maintains this list and how often is it updated?*

+ **[Prevention]** *What measures are in place to prevent IDs from being compromised?*
Some options: MFA (2FA or 3FA), frequent password changes, user education, email or web security that stops attacks based on users clicking on malicious email attachments or web links (URLs).

+ **[Prevention]** *Are we using Machine Identity Management to keep unauthorized devices out? If so, out of what? If not, what would it cost?*

+ **[Prevention]** (For organizations with workloads requiring certificates or devices leveraging certificates for added security): *How effective is our Machine ID management? Do we have problems with expired certificates?*

Workstations and IOT Devices

WS & IOT Devices - The Most Critical Questions

- **[Response]** *If all of our workstations were hit with a Ransomware attack, how quickly could we get back to work?*

- **[Response]** *Ransomware aside, what are our plans to deal with compromised workstations and IOT devices?*

- **[Response]** *Have we tested these plans?*

- **[Prevention]** *What measures do we have in place to ensure the IOT devices on our network aren't able to act as attacker beachheads and disrupt business or steal data?*

- **[Detection]** *How quickly and how often are we detecting compromised workstations or IOT devices?*

- **[Response]** *Is there a long backlog of systems needing remediation?*

Workstations & IOT Devices - Conversation

+ **[Inventory]** *What types of devices do our users use to access our data? Who maintains the list of these devices and how often is the list refreshed?*

    − Desktops − Windows, Mac, or Linux

    − smartphones

    − Tablets

    − IOT devices

+ **[Prevention]** *What is our patching strategy?*

+ **[Prevention and Detection]** *What solutions have we deployed to Prevent and Detect attacks on our workstations?*

+ **[Prevention]** *Are we securing smartphones and tablets with on-device security? Why or why not?*

+ **[Detection]** *Are we using EDR to add to our alerting capabilities?*

+ **[Detection]** *How are we Detecting anomalies in systems where we have no Protection* (IOT devices, for example) *or Detection?* (IOT devices or workstations without EDR) Note: XDR solutions provide this Detection if needed.

+ **[Detection]** *How many workstation alerts (Detections) are we seeing each week (or day)? What about workload alerts?*

+ **[Response]** *Do the alerts overwhelm our ability to Respond and fix the systems? Are we ignoring some alerts?*

Network

Network - The Most Critical Questions

- [**Response**] *What kinds of attacks could cripple our network and what plans do we have in place to deal with them?*

- [**Response**] *Have these plans been tested?*

- [**Response**] *What solutions do we have in place that can stop our sensitive data from leaving our private network? How often have these measures worked and how well have they worked?*

- [**Prevention**] *Who can connect to our wireless (Wi-Fi) or wired networks?*

- [**Detection and Response**] *Are we monitoring the network for unauthorized devices? If we find one, what do we do about it?*

Network - Conversation

- + [**Inventory**] *(As noted above) Do we have an accurate and up-to-date network map? How often does our network change?*

- + [**Prevention**] *Who and what can connect to our network? Using cables or Wi-Fi?*

- + [**Risk**] *If a criminal manages to connect to our network, what can they access?*
  Note: the answers will likely be different for Wi-Fi vs. wired connections.

- + [**Detection**] *If a criminal connected to our network, how quickly would we detect that connection?*
  Note: the answers will likely be different for Wi-Fi vs. wired connections.

- + [**Prevention**] *Which users have network access to which assets?*
  Note: This could be a very long conversation. Ideally, you want assurances that every user cannot access every asset and that high-value assets are only accessible to specific users or systems.

+ [**Prevention**] For networks with remote users – *What assets do VPN users have access to?*
Note: This could also be a very long conversation. Ideally, you want assurances that VPN users cannot access every asset.

Servers or Workloads

Servers or Workloads - The Most Critical Questions

- [**Response**] *If all of our servers were hit with a Ransomware attack, how quickly could we get back to work?*

- [**Response**] *Aside from Ransomware, what are our plans to deal with compromised workloads? How much downtime do these plans include?*

- [**Response**] *Have we tested these plans?*

- [**Prevention & Detection**] *Have we detected any attacks targeting our servers? How did we incorporate info from these attacks into our defenses?*

- [**Prevention**] *What measures do we have in place to ensure the data from our servers is not stolen?*

- [**Prevention and Detection**] (If applicable) *What additional Prevention and Detection measures have we deployed on our public-facing servers? What Response plans do we have that are specifically designed for these servers?*

Servers or Workloads - Conversation

+ [**Inventory**] *Do we have an accurate server count?*
Note: In dynamic environments, particularly those with in-house application development, tracking workloads may be close to impossible.

+ [**Prevention**] If Applicable - *How well are the measures deployed on our public-facing servers working? Do we have metrics on what they've stopped?*

+ [**Detection**] If Applicable - *How many attacks coming through our web servers are we detecting?*

+ [**Prevention**] *How robust is our Machine ID Management solution? How are we stress testing it? Which vendor's solutions are we using or are we relying on a home-made solution?*

+ [**Prevention and Detection**] *Are we protecting these servers with dedicated software or are we treating them like workstations?*

+ [**Detection**] *Are we protecting some or all servers with EDR?*

+ [**Detection**] *How quickly would we detect if a server has been compromised?*

+ [**Detection**] *How many alerts do our servers generate per week (or day)?*

+ [**Response**] *Are we able to address all our server alerts?*

The table below summarizes a Cyber defense review conversation. Ideally, there are tested Response plans for each asset silo. Note that in this example, there are no plans to deal with compromised user identities or Detection measures for identity compromise.

Asset Type	Response	Detection	Prevention
User Identity	No Plan	None	Some MFA
Workstation — IOT	Untested Plans	OK-EDR, no XDR	OK
Network	Tested Plan	OK	OK
Workload	Untested Plan	OK	OK

Breach Post-Mortem

In this hypothetical situation, files from the CFO's laptop were stolen and details on a proposed acquisition were made public, killing the deal. As a business leader, there are four main questions you should ask about a situation like this:

1. *How well did our plans to deal with this eventuality work?*

2. *Do we understand **how** the attackers were able to steal the CFO's information?*
   In other words, do we know how the attackers were able to succeed? What measures did they bypass, what measures did they defeat?

3. *Did any of our solutions **detect** the attack?*
   Prevention failed - that's a given. We want to know if Detection worked. Did one of the Detection solutions alert us about the issue? If so, did we then fail to see it or act on it? What can we do to improve the process to detect this type of attack in the future? If we didn't detect it, what can we do to increase our detection coverage?

4. *What are we doing to stop this type of attack in the future?*
   The response should be framed in terms of changes to Prevention, Detection, and Response.

   > **Prevention Example:** *The latest patch prevents this kind of attack from succeeding in the future. We will have the patch deployed to all of our systems by Friday.*

   > **Detection Example:** *Our new XDR solution detects this kind of attack. It is already in place.*

   > **Response Example:** *We were alerted and could have stopped the attack before the data was stolen, but we missed it. We are reviewing our alert prioritization procedures to make sure these alerts are not ignored in the future.*

7

Author's Postscript

I wrote this book to help business leaders better understand the Cyber defenses some of them are asked to review, approve, and fund. I believe adding their informed voices to the Cybersecurity conversation will improve Information Security.

I also wanted to provide business leaders with enough information to have meaningful and productive conversations with Infosec leaders. If you feel you need expert help, or someone to act as an intermediary and facilitate these conversations, or if you have feedback about the ideas in this book, contact the author at **advisor@fieldstarglobal.com**.

Thank you.

Mike Gable
September 2021

Thank You

Thanks to everyone who helped make this possible, especially:

- Josh Borges

- Stuart Jones

- Tom Miller

- James Pulley

- Lorri Schoeni

- Jeff Stewart

- Curt Walleen

Index

2FA, 22, 23
3FA, 22, 23

administrator, 32
adversaries, 7
Alert Fatigue, 14, 17
anti-virus, 27, 43
attacker, 4
AV, 27

biometric, 22
breach, 5, 16, 32, 42

CAE, 4
CEO, 4, 27
certificate, 20, 26, 27, 40, 42
CFO, 4
CISO, 5
cloud, 4, 30
concentrator, 35, 37
container, 39
credential, 20
CRM, 38
crown jewels, 17
CSO, 5
cyber, 3, 4, 45
cyberattack, 4
cybercriminal, 5, 30
cybercurrency, 6
cybersecurity, v, 1–5, 20, 27, 43

DDOS, 5
defenses, 7
detection, 2, 9, 10, 13, 21, 25, 27, 39
disruption, 6
drone, 24

EDR, 25, 39
email, 30, 38
EPP, 27

Eternal Blue, 28, 29
EUBA, 22
extortion, 6

Finance, 33
fingerprint, 22
firewall, 30, 32

hijack, 35
HR, 33, 35

IAM, 21
identity, 19, 20
identity theft, 20
IGA, 21
Infosec, 1, 4, 5
infosec, 19, 28, 32, 37, 43
IOT, 19, 24–26, 28, 29, 35, 40, 43
IRT, 11

Legal, 33
Linux, 43

machine identity management, 27, 42
Macintosh, 23
malware, 4, 15, 24
McAfee, 27
MFA, 22
Microsoft, 27–30

network, 4, 19, 27, 29, 31, 32
neurosurgery, 4
NSA, 28, 29
NTA, 25

Oracle, 30

PAM, 22
pandemic, 37
password, 22

www.ingramcontent.com/pod-product-compliance
Lightning Source LLC
Chambersburg PA
CBHW041209220326
41597CB00030BA/5250